INDUCTOR BASICS

FIRST EDITION

BY PRASUN BARUA

ABOUT

Welcome to Inductor Basics! This is a nonfiction science book which contains various topics on basics of inductor. When an electric current flows through the inductor's coil, it momentarily stores energy in a magnetic field. An inductor is made up of two terminals and an insulated wire coil that either loops around air or around a core substance that boosts the magnetic field. Inductors aid to accommodate fluctuations in an electric current going through a circuit. When an electric current flows through a conductor like copper wire, it creates a small magnetic field surrounding the wire. The magnetic field grows substantially stronger when the wire is formed into a coil. When the wire is wound around a central core composed of a substance like iron, the magnetic field becomes even stronger; this is essentially how an electromagnet works. The magnetic field is entirely determined by the electric current. Inductors adjust for changes in current flow by using the relationship between the electric current and the magnetic field. When current flows through the inductor's coil, the magnetic field expands until it ultimately stabilizes. Until then, the coil prevents current flow. After the magnetic field has stabilized, the current flows normally through the coil. As long as the current flows through the coil, energy is stored in the magnetic field. When the current is turned off, the magnetic field begins to collapse, and the magnetic energy is converted back into electrical energy, which flows back into the circuit until the magnetic field entirely collapses. Thanks for reading the book.

CONTENTS

CHAPTER-1: WHAT IS INDUCTOR?

An inductor is a passive electrical component made of a coil of wire that is made to benefit from the interaction between electricity and magnetism when an electric current flows through the coil. Even a straight piece of conductive wire can include some inductance. An inductor is an electrical component used to induce inductance into a circuit, which opposes changes in current flow, both in magnitude and direction. We know that a magnetic flux forms around a wire conductor when an electrical current passes through it.

Due to this effect, there is a correlation between the direction of the current flowing through a conductor and the direction of the magnetic flux that is revolving around it. As a result, "Fleming's Right Hand Rule," a correlation between current and magnetic flux direction, is created.

The movement of the magnetic flux, which opposes or resists any changes in the electrical current flowing through it, induces a secondary voltage into the same coil, which is another significant property of a coiled coil.

A Typical Inductor

An inductor is nothing more than a coil of wire twisted around a central core in its most basic form.

For the majority of coils, the magnetic flux (NΦ) produced by the current (I) flowing through the coil is proportional to this flow of electrical current.

Another passive type of electrical component, an inductor (also known as a choke), is made of a coil of wire that is intended to take advantage of this relationship by creating a magnetic field either within the coil itself or within its core as a result of the current passing through the coil of wire.

A wire coil made into an inductor creates a magnetic field that is far greater than the magnetic field created by a plain wire coil. To concentrate their magnetic flux, inductors are made by tightly wrapping wire around a solid central core, which can be either a straight cylindrical rod or a continuous loop or ring.

Since a coil of wire serves as the schematic representation of an inductor, a coil of wire may also be referred to as an inductor. Typically, inductors are divided into groups based on the type of inner core they are wound around, such as a hollow core (free air), solid iron core, or soft ferrite core. The various core types are distinguished by adding continuous or dotted parallel lines next to the wire coil, as is illustrated below.

Inductor Symbol

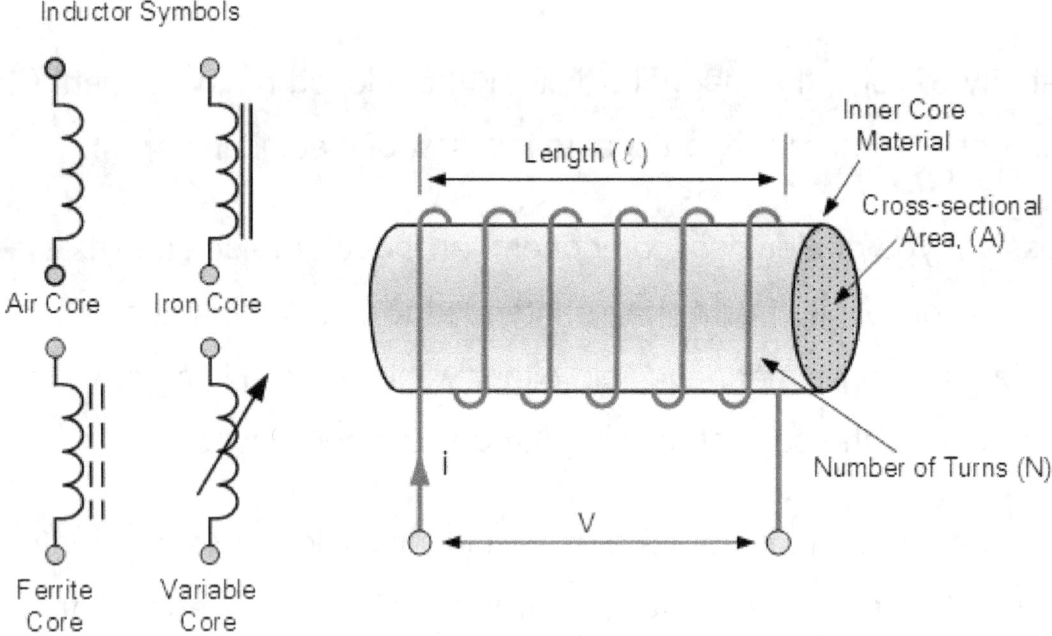

A magnetic flux produced by an inductor is proportional to the current, I that travels through it. Contrary to capacitors, which resist changes in voltage across their plates, inductors resist changes in the rate at which current flows through them because of the accumulation of self-induced energy inside their magnetic fields.

To put it another way, inductors will readily transmit a steady state DC current but will oppose or resist variations in current. It is known as inductance and is denoted by the symbol L with Henry (H) units in honor of Joseph Henry. Inductance is the ability of an inductor to resist variations in current and which also relates current, I with its magnetic flux linkage, N, as a constant of proportionality. Sub-units of the Henry are employed to represent its value for smaller inductors because the Henry is a relatively big unit of inductance in and of itself.

Inductance Prefixes

Prefix	Symbol	Multiplier	Power of Ten
milli	m	1/1,000	10^{-3}
micro	µ	1/1,000,000	10^{-6}
nano	n	1/1,000,000,000	10^{-9}

In order to display the sub-units of the Henry we would use as an example:

- 1mH = 1 milli-Henry – which is equal to one thousandths (1/1000) of an Henry.
- 100µH = 100 micro-Henries – which is equal to 100 millionth's (1/1,000,000) of a Henry.

The shape of the coil, the number of turns of the insulated wire, the number of layers of wire, the distance between the turns, the permeability of the core material, the size or cross-sectional area of the core, etc. are just a few of the many factors that affect a coil's inductance in electrical circuits.

A central core area (A) and a fixed number of wire turns per unit length define an inductor coil (I). Therefore, every current I that travels through a coil of N turns that is coupled by an amount of magnetic flux has a flux linkage of N and will generate a magnetic flux in the opposite direction of the current flow. Then, in accordance with Faraday's Law, any modification to this magnetic flux connection results in a self-induced voltage of:

$$V_L = N\frac{d\Phi}{dt} = \frac{\mu N^2 A}{\ell}\frac{di}{dt}$$

- Where:
- N is the number of turns
- A is the cross-sectional Area in m^2
- Φ is the amount of flux in Weber
- μ is the Permeability of the core material
- l is the Length of the coil in meters
- di/dt is the Currents rate of change in amps/second

A time varying magnetic field induces a voltage that is proportional to the rate of change of the current producing it with a positive value indicating an increase in emf and a negative value indicating a decrease in emf. The equation relating this self-induced voltage, current and inductance can be found by substituting the $\mu N^2 A\,/\,l$ with L denoting the constant of proportionality called the Inductance of the coil.

The relation between the flux in the inductor and the current flowing through the inductor is given as: $N\Phi = Li$. As an inductor consists of a coil of conducting wire, this then reduces the above equation to give the self-induced emf, sometimes called the back emf induced in the coil too:

Back emf Generated by an Inductor

$$V_L(t) = \frac{d\phi}{dt} = \frac{d\,Li}{dt} = -L\frac{di}{dt}$$

Where: L is the self-inductance and di/dt the rate of current change.

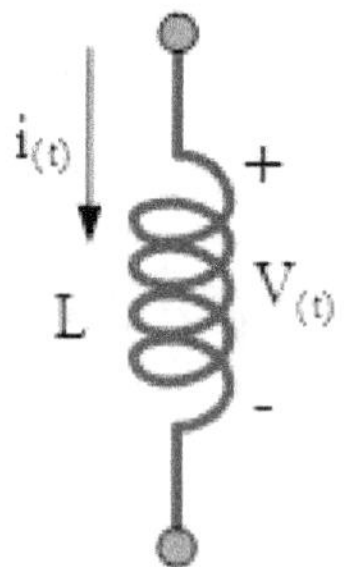

Inductor Coil

A circuit with an inductance of one Henry will induce an emf of one volt when the current running through the circuit varies at a rate of one ampere per second, according to this equation, which states that "Self-induced emf equals Inductance times the rate of current change."

There is an essential detail to remember about the equation above. It only links the induced emf voltage to changes in current since the induced emf voltage will be zero if the inductor current is constant and not changing, as it would be in a steady-state DC current. This is because the instantaneous rate of current change is zero, or $di/dt = 0$. With a steady state DC current flowing through the inductor and therefore zero induced voltage across it, the inductor acts as a short circuit equal to a piece of wire, or at the very least a very low value resistance. In other words, the opposition to the flow of current offered by an inductor is very different between AC and DC circuits.

The Time Constant of an Inductor

We know that an inductor's current cannot change instantly because, in order for this to happen, the current would have to change by a finite amount in zero time, causing the current change rate to be infinite ($di/dt =$

∞), which would result in an infinite induced emf and the absence of infinite voltages. High voltages can, however, be induced across an inductor's coil if the current through it fluctuates dramatically, as it does when a switch is turned on.

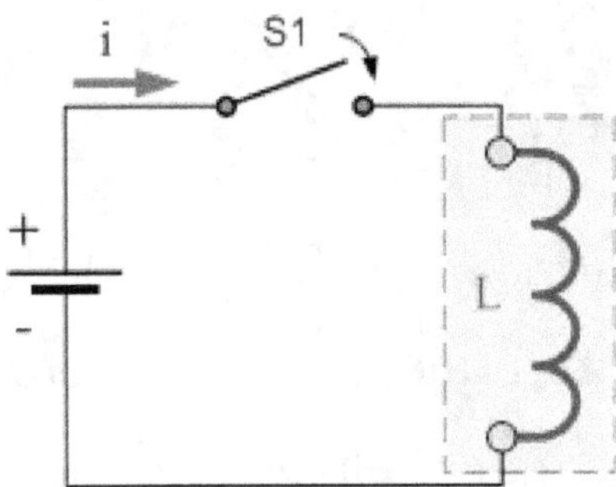

Consider the circuit of a pure inductor on the right. With the switch, (S1) open, no current flows through the inductor coil. As no current flows through the inductor, the rate of change of current (di/dt) in the coil will be zero. If the rate of change of current is zero there is no self-induced back-emf, ($V_L = 0$) within the inductor coil.

If we now close the switch (t = 0), a current will flow through the circuit and slowly rise to its maximum value at a rate determined by the inductance of the inductor. This rate of current flowing through the inductor multiplied by the inductors inductance in Henry's, results in some fixed value self-induced emf being produced across the coil as determined by Faraday's equation above, $V_L = -Ldi/dt$.

This self-induced emf across the inductors coil, (V_L) fights against the applied voltage until the current reaches its maximum value and a steady state condition is reached. The current which now flows through the coil is determined only by the DC or "pure" resistance of the coils windings as the reactance value of the coil has decreased to zero because the rate of change of current (di/dt) is zero in a steady state condition. In other words, in a real coil only the coils DC resistance exists to oppose the flow of current through itself.

Likewise, if switch (S1) is opened, the current flowing through the coil will start to fall but the inductor will again fight against this change and try to keep the current flowing at its previous value by inducing another voltage in the other direction. The slope of the fall will be negative and related to the inductance of the coil as shown below.

Current and Voltage in an Inductor

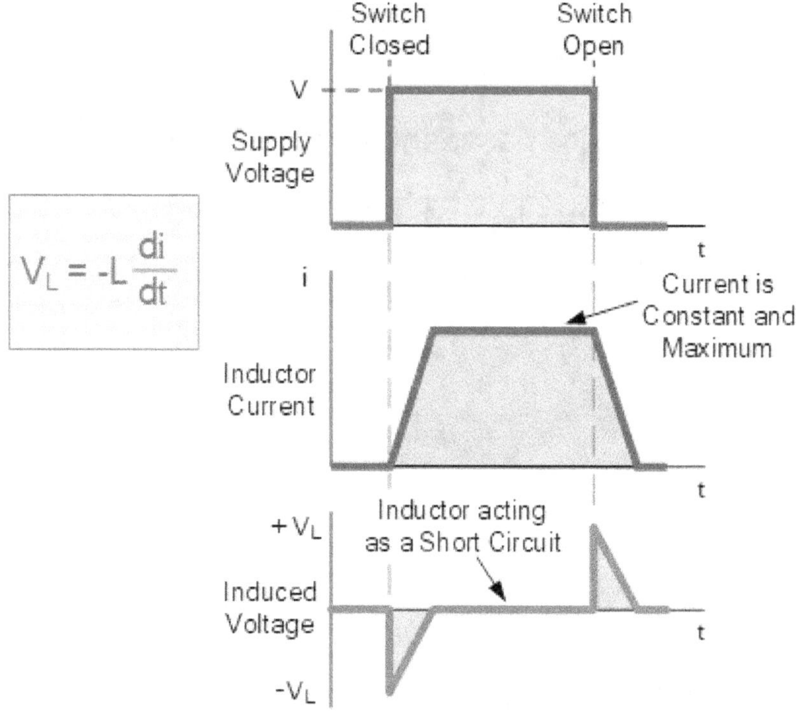

Depending on the rate of current change, the inductor will produce a certain amount of induced voltage. The direction of an induced electromagnetic field is such that it will always oppose the change that is creating it, according to Lenz's Law, which we discussed in our section on electromagnetic induction.

To put it another way, an induced emf always opposes the motion or change that caused it to begin with. As a result, when the current is dropping, the voltage polarity will operate as a source, and when the current is growing, it will act as a load. Therefore, the amount of the induced emf will increase or decrease at the same rate of current change via the coil.

Example No1

A steady state direct current of 4 ampere passes through a solenoid coil of 0.5H. What would be the average back emf voltage induced in the coil if the switch in the above circuit was opened for 10mS and the current flowing through the coil dropped to zero ampere.

$$V_L = L\frac{di}{dt} = 0.5\frac{4}{0.01} = 200\,volts$$

Power in an Inductor

We know that an inductor in a circuit opposes the flow of current, (i) through it because the flow of this current induces an emf that opposes it, Lenz's Law. Then work has to be done by the external battery source in order to keep the current flowing against this induced emf. The instantaneous power used in forcing the current, (i) against this self-induced emf, (V_L) is given from above as:

$$V_{L(t)} = -L \frac{di}{dt}$$

Power in a circuit is given as, P = V*I therefore:

$$P = v.i = \left(L \frac{di}{dt} \right) \times i = \frac{1}{2} L \frac{di^2}{dt} = \frac{d}{dt} \left[\frac{1}{2} L i^2 \right]$$

An ideal inductor has no resistance only inductance so $R = 0 \; \Omega$ and therefore no power is dissipated within the coil, so we can say that an ideal inductor has zero power loss.

The Energy Stored

When power flows into an inductor, energy is stored in its magnetic field. When the current flowing through the inductor is increasing and di/dt becomes greater than zero, the instantaneous power in the circuit must also be greater than zero, (P > 0) ie, positive which means that energy is being stored in the inductor.

Likewise, if the current through the inductor is decreasing and di/dt is less than zero then the instantaneous power must also be less than zero, (P < 0) ie, negative which means that the inductor is returning energy back into the circuit. Then by integrating the equation for power above, the total magnetic energy which is always positive, being stored in the inductor is therefore given as:

Energy Stored

$$W_{(t)} = \frac{1}{2} L i^2_{(t)}$$

Where: W is in joules, L is in Henries and i is in Amperes

The energy is actually being stored within the magnetic field that surrounds the inductor by the current flowing through it. In an ideal inductor that has no resistance or capacitance, as the current increases energy flows into the inductor and is stored there within its magnetic field without loss, it is not released until the current decreases and the magnetic field collapses.

Then in an alternating current, AC circuit an inductor is constantly storing and delivering energy on each and every cycle. If the current flowing through the inductor is constant as in a DC circuit, then there is no change in the stored energy as $P = Li(di/dt) = 0$.

Therefore, inductors can be defined as passive components as they can both stored and deliver energy to the circuit, but they cannot generate energy. An ideal inductor is classed as loss less, meaning that it can store energy indefinitely as no energy is lost.

However, real inductors will always have some resistance associated with the windings of the coil and whenever current flows through a resistance energy is lost in the form of heat due to Ohms Law, ($P = I^2 R$) regardless of whether the current is alternating or constant.

Then the primary use for inductors is in filtering circuits, resonance circuits and for current limiting. An inductor can be used in circuits to block or reshape alternating current or a range of sinusoidal frequencies, and in this role an inductor can be used to "tune" a simple radio receiver or various types of oscillators. It can also protect sensitive equipment from destructive voltage spikes and high inrush currents.

CHAPTER-2: INDUCTANCE OF A COIL

Inductance is the property of a component that resists the change in current passing through it, and even a straight piece of wire has some inductance. The electrical property of an inductive coil to oppose any change in the current running through it is referred to as its inductance.

As a result, inductance is only present in an electric circuit when the current changes. As a result of their changing magnetic field, inductors generate a self-induced emf within themselves.

In an electrical circuit, when the emf is induced in the same circuit where the current is changing, this effect is known as self-induction (L), but it is also known as back-emf because its polarity is opposite to the applied voltage.

When the emf is induced into an adjacent component situated within the same magnetic field, the emf is said to be induced by Mutual-induction, (M) and mutual induction is the basic operating principal of transformers, motors, relays etc. Self-inductance is a special case of mutual inductance, and because it is produced within a single isolated circuit we generally call self-inductance simply, Inductance.

The basic unit of measurement for inductance is called the Henry, (H) after Joseph Henry, but it also has the units of Webers per Ampere (1 H = 1 Wb/A).

Lenz's Law tells us that an induced emf generates a current in a direction which opposes the change in flux which caused the emf in the first place, the principal of action and reaction. Then we can accurately define Inductance as being: "a coil will have an inductance value of one Henry when an emf of one volt is induced in the coil were the current flowing through the said coil changes at a rate of one ampere/second".

In other words, a coil has an inductance, (L) of one Henry, (1H) when the current flowing through the coil changes at a rate of one ampere/second, (A/s). This change induces a voltage of one volt, (V_L) in it. Thus the mathematical representation of the rate of change of current through a wound coil per unit time is given as:

$$\frac{di}{dt} \quad (A/s)$$

Where: *di* is the change in the current in Amperes and *dt* is the time taken for this current to change in seconds. Then the voltage induced in a coil, (V_L) with an inductance of L Henries as a result of this change in current is expressed as:

$$V_L = -L\frac{di}{dt} \quad (V)$$

It should be noted here that the negative sign indicates that voltage induced opposes the change in current through the coil per unit time (di/dt).

From the above equation, the inductance of a coil can therefore be presented as:

Inductance of a Coil

$$L = \frac{V_L}{(di/dt)} = \frac{1\,volt}{1A/s} = 1\,Henry$$

Where: L is the inductance in Henries, V_L is the voltage across the coil and di/dt is the rate of change of current in Amperes per second, A/s.

Inductance, L is actually a measure of an inductors "resistance" to the change of the current flowing through the circuit and the larger is its value in Henries, the lower will be the rate of current change.

We know that inductors are devices that can store their energy in the form of a magnetic field. Inductors are made from individual loops of wire combined to produce a coil and if the number of loops within the coil are increased, then for the same amount of current flowing through the coil, the magnetic flux will also increase.

Therefore, by increasing the number of loops or turns within a coil, increases the coils inductance. Then the relationship between self-inductance, (L) and the number of turns, (N) and for a simple single layered coil can be given as:

Self-Inductance of a Coil

$$L = N\frac{\Phi}{I}$$

- Where:
-     L is in Henries
-     N is the Number of Turns
-     Φ is the Magnetic Flux
-     I is in Amperes

This expression can also be defined as the magnetic flux linkage, ($N\Phi$) divided by the current, as effectively the same value of current flows through each turn of the coil. Note that this equation only applies to linear magnetic materials.

Inductance Example No1

A hollow air cored inductor coil consists of 500 turns of copper wire which produces a magnetic flux of 10mWb when passing a DC current of 10 amps. Calculate the self-inductance of the coil in milli-Henries.

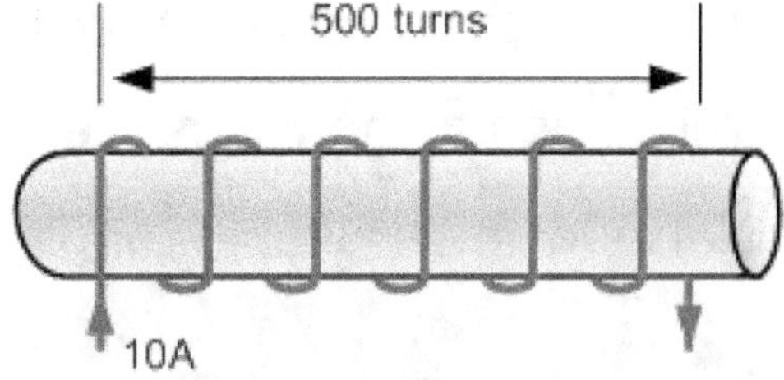

$$L = N\frac{\Phi}{I} = 500\frac{0.01}{10} = 500\text{mH}$$

Inductance Example No2

Calculate the value of the self-induced emf produced in the same coil after a time period of 10 milli-seconds (10ms).

$$\text{emf} = L\frac{di}{dt} = 0.5\frac{10}{0.01} = 500\text{V}$$

The self-inductance of a coil or to be more precise, the coefficient of self-inductance also depends upon the characteristics of its construction. For example, size, length, number of turns etc. It is therefore possible to have inductors with very high coefficients of self-induction by using cores of a high permeability and a large number of coil turns. Then for a coil, the magnetic flux that is produced in its inner core is equal to:

$$\Phi = B.A$$

Where: Φ is the magnetic flux, B is the flux density, and A is the area.

If the inner core of a long solenoid coil with N number of turns per metre length is hollow, "air cored", then the magnetic induction within its core will be given as:

$$B = \mu_o H = \mu_o \frac{N.I}{\ell}$$

Then by substituting these expressions in the first equation above for Inductance will give us:

$$L = N\frac{\Phi}{I} = N\frac{B.A}{I} = N\frac{\mu_o.N.I}{\ell.I}.A$$

By cancelling out and grouping together like terms, then the final equation for the coefficient of self-inductance for an air cored coil (solenoid) is given as:

$$L = \mu_o \frac{N^2.A}{\ell}$$

- Where:
-     L is in Henries
-     μ_o is the Permeability of Free Space ($4.\pi.10^{-7}$)
-     N is the Number of turns
-     A is the Inner Core Area (πr^2) in m^2
-     ℓ is the length of the Coil in meters

Since the inductance of a coil is due to the magnetic flux around it, the stronger the magnetic flux for a given value of current the greater will be the inductance. Therefore, a coil of many turns will have a higher inductance value than one of only a few turns and therefore, the equation above will give inductance L as being proportional to the number of turns squared N^2.

Not only increasing the number of coils turns, but also, we can also increase inductance by increasing the coils diameter or making the core longer. In both cases more wire is required to construct the coil and therefore, more lines of force exist to produce the required back emf. The inductance of a coil can be increased further still if the coil is wound onto a ferromagnetic core, that is one made of a soft iron material, then one wound onto a non-ferromagnetic or hollow air core.

Ferrite Core

If the inner core is made of some ferromagnetic material such as soft iron, cobalt or nickel, the inductance of the coil would greatly increase because for the same amount of current flow the magnetic flux generated would be much stronger. This is because the material concentrates the lines of force more strongly through the softer ferromagnetic core material as we saw in the Electromagnets.

As an example, if the core material has a relative permeability 1000 times greater than free space, $1000\mu_o$ such as soft iron or steel, then the inductance of the coil would be 1000 times greater so we can say that the inductance of a coil increases proportionally as the permeability of the core increases.

Then for a coil wound around a former or core the inductance equation above would need to be modified to include the relative permeability μ_r of the new former material.

If the coil is wound onto a ferromagnetic core a greater inductance will result as the cores permeability will change with the flux density. However, depending upon the type of ferromagnetic material, the inner cores magnetic flux may quickly reach saturation producing a non-linear inductance value. Since the flux density around a coil of wire depends upon the current flowing through it, inductance, L also becomes a function of this current flow, i.

CHAPTER-3: MUTUAL INDUCTANCE

The effect of one coil's magnetic field on another coil as it produces a voltage in the adjacent coil is referred to as mutual inductance. Mutual inductance is a circuit characteristic that determines the ratio of a time-varying magnetic flux generated by one coil being induced into an adjacent second coil. We know that an inductor generates a self-induced emf as a result of a changing magnetic field around its own coil turns. When this emf is induced in the same circuit in which the current is changing this effect is called Self-induction (L).

However, when the emf is induced into an adjacent coil situated within the same magnetic field, the emf is said to be induced magnetically, inductively or by Mutual induction, symbol (M). Then when two or more coils are magnetically linked together by a common magnetic flux, they are said to have the property of Mutual Inductance.

Mutual Inductance is the basic operating principal of the transformer, motors, generators and any other electrical component that interacts with another magnetic field. Then we can define mutual induction as the current flowing in one coil that induces a voltage in an adjacent coil.

But mutual inductance can also be a bad thing as "stray" or "leakage" inductance from a coil can interfere with the operation of another adjacent component by means of electromagnetic induction, so some form of electrical screening to a ground potential may be required.

The amount of mutual inductance that links one coil to another depends very much on the relative positioning of the two coils. If one coil is positioned next to the other coil so that their physical distance apart is small, then nearly all of the magnetic flux generated by the first coil will interact with the coil turns of the second coil inducing a relatively large emf and therefore producing a large mutual inductance value.

Similarly, if the two coils are farther apart from each other or at different angles, the amount of induced magnetic flux from the first coil into the second will be weaker producing a much smaller induced emf and therefore a much smaller mutual inductance value. So the effect of mutual inductance is very much dependent upon the relative positions or spacing, (S) of the two coils and this is demonstrated below.

Mutual Inductance between Coils

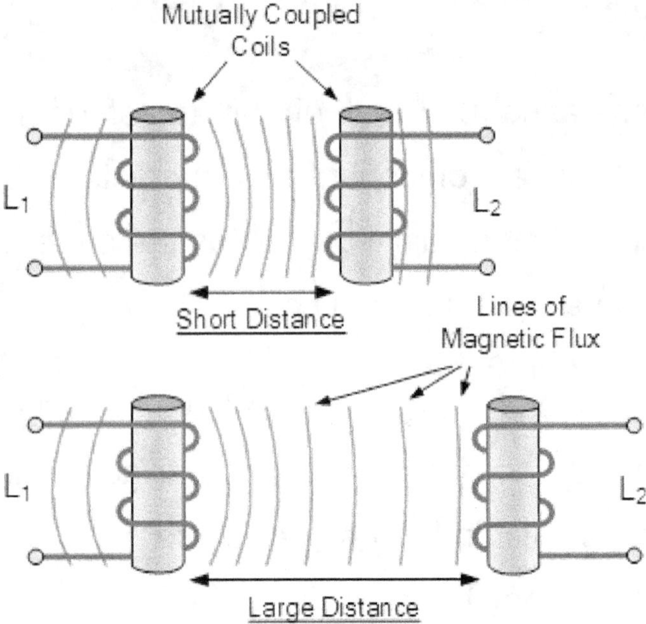

The mutual inductance that exists between the two coils can be greatly increased by positioning them on a common soft iron core or by increasing the number of turns of either coil as would be found in a transformer.

If the two coils are tightly wound one on top of the other over a common soft iron core unity coupling is said to exist between them as any losses due to the leakage of flux will be extremely small. Then assuming a perfect flux linkage between the two coils the mutual inductance that exists between them can be given as.

$$M = \frac{\mu_0 \mu_r N_1 N_2 A}{\ell}$$

Where:

- μ_o is the permeability of free space ($4.\pi.10^{-7}$)
- μ_r is the relative permeability of the soft iron core
- N is in the number of coils turns
- A is in the cross-sectional area in m^2
- ℓ is the coils length in meters

Mutual Induction

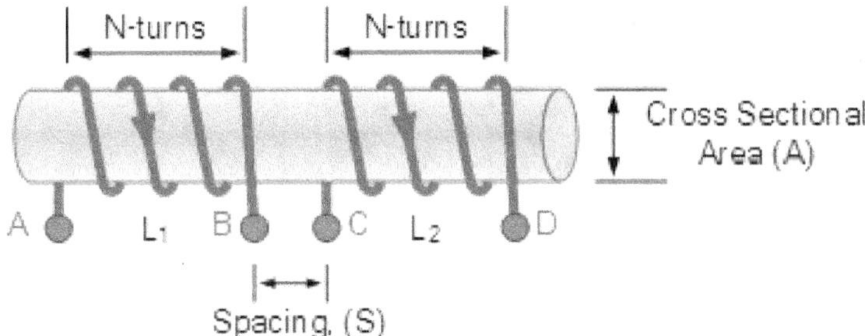

In this case, the current flowing in coil one, L_1 sets up a magnetic field around itself with some of these magnetic field lines passing through coil two, L_2 giving us mutual inductance. Coil one has a current of I_1 and N_1 turns while, coil two has N_2 turns. Therefore, the mutual inductance, M_{12} of coil two that exists with respect to coil one depends on their position with respect to each other and is given as:

$$M_{12} = \frac{N_2 \Phi_{12}}{I_1}$$

Similarly, the flux linking coil one, L_1 when a current flows around coil two, L_2 is exactly the same as the flux linking coil two when the same current flows around coil one above, then the mutual inductance of coil one with respect of coil two is defined as M_{21}. This mutual inductance is true irrespective of the size, number of turns, relative position or orientation of the two coils. Because of this, we can write the mutual inductance between the two coils as: $M_{12} = M_{21} = M$.

It has been observed that self-inductance characterizes an inductor as a single circuit element, while mutual inductance signifies some form of magnetic coupling between two inductors or coils, depending on their distance and arrangement. The self-inductance of each individual coil is given as:

$$L_1 = \frac{\mu_0 \mu_r N_1^2 A}{\ell} \quad \text{and} \quad L_2 = \frac{\mu_0 \mu_r N_2^2 A}{\ell}$$

By cross-multiplying the two equations above, the mutual inductance, M that exists between the two coils can be expressed in terms of the self-inductance of each coil.

$$M^2 = L_1 L_2$$

giving us a final and more common expression for the mutual inductance between the two coils of:

Mutual Inductance Between Coils

$$M = \sqrt{L_1 L_2} \ \ H$$

However, the above equation assumes zero flux leakage and 100% magnetic coupling between the two coils, L_1 and L_2. In reality there will always be some loss due to leakage and position, so the magnetic coupling between the two coils can never reach or exceed 100%, but can become very close to this value in some special inductive coils.

If some of the total magnetic flux links with the two coils, this amount of flux linkage can be defined as a fraction of the total possible flux linkage between the coils. This fractional value is called the coefficient of coupling and is given the letter k.

Coupling Coefficient

Generally, the amount of inductive coupling that exists between the two coils is expressed as a fractional number between 0 and 1 instead of a percentage (%) value, where 0 indicates zero or no inductive coupling, and 1 indicating full or maximum inductive coupling.

In other words, if k = 1 the two coils are perfectly coupled, if k > 0.5 the two coils are said to be tightly coupled and if k < 0.5 the two coils are said to be loosely coupled. Then the equation above which assumes a perfect coupling can be modified to take into account this coefficient of coupling, k and is given as:

Coupling Factor Between Coils

$$k = \frac{M}{\sqrt{L_1 L_2}} \quad \text{or} \quad M = k\sqrt{L_1 L_2}$$

When the coefficient of coupling, k is equal to 1, (unity) such that all the lines of flux of one coil cuts all of the turns of the second coil, that is the two coils are tightly coupled together, the resulting mutual inductance will be equal to the geometric mean of the two individual inductances of the coils.

Furthermore, when the inductances of the two coils are the same and equal, L_1 is equal to L_2, the mutual inductance that exists between the two coils will equal the value of one single coil as the square root of two equal values is the same as one single value as shown.

$$M = \sqrt{L_1 L_2} = L$$

Mutual Inductance Example No1

Two inductors whose self-inductances are given as 75mH and 55mH respectively, are positioned next to each other on a common magnetic core so that 75% of the lines of flux from the first coil are cutting the second coil. Calculate the total mutual inductance that exists between the two coils.

$$M = k\sqrt{L_1 L_2}$$

$$M = 0.75\sqrt{75mH \times 55mH} = 48.2mH$$

Mutual Inductance Example No2

When two coils having inductances of 5H and 4H respectively were wound uniformly onto a non-magnetic core, it was found that their mutual inductance was 1.5H. Calculate the coupling coefficient that exists between.

$$k = \frac{M}{\sqrt{L_1 L_2}} = \frac{1.5}{\sqrt{5 \times 4}} = 0.335 = 33.5\%$$

CHAPTER-4: INDUCTORS IN SERIES

When inductors are daisy chained together and share an electrical current, they can be connected in a series connection. These connections between inductors result in increasingly intricate networks whose total inductance is the sum of their individual inductors.

To connect inductors in series or parallel, there are a few rules that are based on the idea that there is no magnetic coupling or mutual inductance between the individual inductors. When inductors are daisy-chained together in a straight line, end to end, they are said to be connected in "Series."

We are aware that the various resistance levels connected in series simply "add" to one another, and the same is true of inductance.

Since the number of coils turns increases when inductors are connected in series, they are merely "added together," with the total circuit inductance LT being equal to the sum of all the individual inductances added together.

Inductor in Series Circuit

The current, (I) that flows through the first inductor, L_1 has no other way to go but pass through the second inductor and the third and so on. Then, series inductors have a Common Current flowing through them, for example:

$I_{L1} = I_{L2} = I_{L3} = I_{AB}$ …etc.

In the example above, the inductors L_1, L_2 and L_3 are all connected together in series between points A and B. The sum of the individual voltage drops across each inductor can be found using Kirchhoff's Voltage Law (KVL) where, $V_T = V_1 + V_2 + V_3$ and we know that inductance that the self-induced emf across an inductor is given as: $V = L\,di/dt$.

Therefore, by taking the values of the individual voltage drops across each inductor in our example above, the total inductance for the series combination is given as:

$$L_T \frac{di}{dt} = L_1 \frac{di}{dt} + L_2 \frac{di}{dt} + L_3 \frac{di}{dt}$$

By dividing through the above equation by di/dt we can reduce it to give a final expression for calculating the total inductance of a circuit when connecting inductors together in series and this is given as:

Inductors in Series Equation

$L_{total} = L_1 + L_2 + L_3 + \ldots + L_n$ etc.

Then the total inductance of the series chain can be found by simply adding together the individual inductances of the inductors in series just like adding together resistors in series. However, the above equation only holds true when there is "NO" mutual inductance or magnetic coupling between two or more of the inductors, (they are magnetically isolated from each other).

One important point to remember about inductors in series circuits, the total inductance (L_T) of any two or more inductors connected together in series will always be GREATER than the value of the largest inductor in the series chain.

Inductors in Series Example No1

Three inductors of 10mH, 40mH and 50mH are connected together in a series combination with no mutual inductance between them. Calculate the total inductance of the series combination.

$$L_T = L_1 + L_2 + L_3 = 10\text{mH} + 40\text{mH} + 50\text{mH} = 100\text{mH}$$

Mutually Connected Inductors in Series

When inductors are connected together in series so that the magnetic field of one links with the other, the effect of mutual inductance either increases or decreases the total inductance depending upon the amount of magnetic coupling. The effect of this mutual inductance depends upon the distance apart of the coils and their orientation to each other.

Mutually connected series inductors can be classed as either "Aiding" or "Opposing" the total inductance. If the magnetic flux produced by the current flows through the coils in the same direction, then the coils are said to be Cumulatively Coupled. If the current flows through the coils in opposite directions, then the coils are said to be Differentially Coupled as shown below.

Cumulatively Coupled Series Inductors

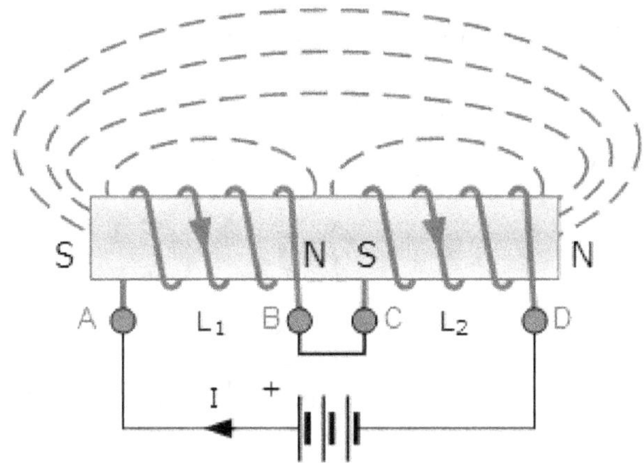

While the current flowing between points A and D through the two cumulatively coupled coils is in the same direction, the equation above for the voltage drops across each of the coils needs to be modified to take into account the interaction between the two coils due to the effect of mutual inductance. The self-inductance of each individual coil, L_1 and L_2 respectively will be the same as before but with the addition of M denoting the mutual inductance.

Then the total emf induced into the cumulatively coupled coils is given as:

$$L_T \frac{di}{dt} = L_1 \frac{di}{dt} + L_2 \frac{di}{dt} + 2\left(M \frac{di}{dt}\right)$$

Where: 2M represents the influence of coil L_1 on L_2 and likewise coil L_2 on L_1.

By dividing through the above equation by di/dt we can reduce it to give a final expression for calculating the total inductance of a circuit when the inductors are cumulatively connected and this is given as:

$$L_{total} = L_1 + L_2 + 2M$$

If one of the coils is reversed so that the same current flows through each coil but in opposite directions, the mutual inductance, M that exists between the two coils will have a cancelling effect on each coil as shown below.

Differentially Coupled Series Inductors

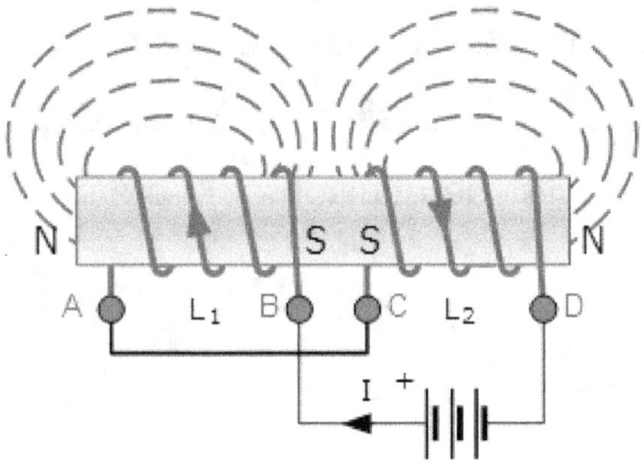

The emf that is induced into coil 1 by the effect of the mutual inductance of coil two is in opposition to the self-induced emf in coil one as now the same current passes through each coil in opposite directions. To take account of this cancelling effect a minus sign is used with M when the magnetic field of the two coils are differentially connected giving us the final equation for calculating the total inductance of a circuit when the inductors are differentially connected as:

$$L_{total} = L_1 + L_2 - 2M$$

Then the final equation for inductively coupled inductors in series is given as:

$$L_T = L_1 + L_2 \pm 2M$$

Example No2

Two inductors of 10mH respectively are connected together in a series combination so that their magnetic fields aid each other giving cumulative coupling. Their mutual inductance is given as 5mH. Calculate the total inductance of the series combination.

$$L_T = L_1 + L_2 + 2M$$
$$L_T = 10mH + 10mH + 2(5mH)$$
$$L_T = 30mH$$

Example No3

Two coils connected in series have a self-inductance of 20mH and 60mH respectively. The total inductance of the combination was found to be 100mH. Determine the amount of mutual inductance that exists between the two coils assuming that they are aiding each other.

$$L_T = L_1 + L_2 \pm 2M$$

$$100 = 20 + 60 + 2M$$

$$2M = 100 - 20 - 60$$

$$\therefore M = \frac{20}{2} = 10\text{mH}$$

In a Brief

By connecting inductors in series, which is equivalent to connecting resistors in series, we may create a total inductance value, LT, equal to the sum of the individual values. However, mutual inductance can have an impact on inductors when they are connected together. Depending on whether the coils are cumulatively linked (in the same direction) or differentially coupled, mutually connected series inductors are categorized as either "aiding" or "opposing" the total inductance (in opposite direction).

CHAPTER-5: INDUCTORS IN PARALLEL

Inductors are said to be connected together in Parallel when both of their terminals are respectively connected to each terminal of another inductor or inductors

The voltage drop across all of the inductors in parallel will be the same. Then, Inductors in Parallel have a Common Voltage across them and in our example below the voltage across the inductors is given as:

$$V_{L1} = V_{L2} = V_{L3} = V_{AB} \text{ ...etc}$$

In the following circuit the inductors L_1, L_2 and L_3 are all connected together in parallel between the two points A and B.

Inductors in Parallel Circuit

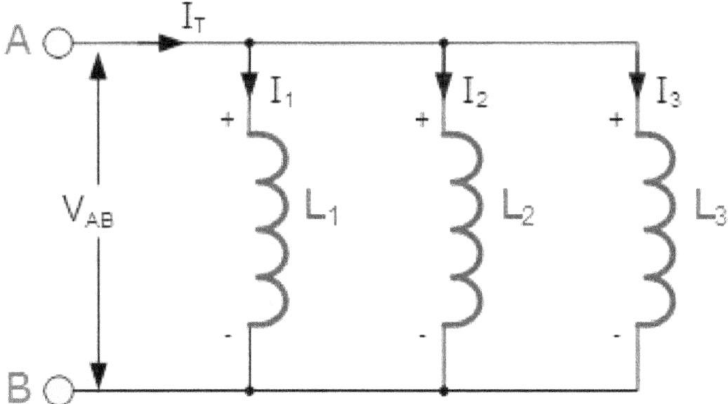

We know that the total inductance, L_T of the circuit was equal to the sum of all the individual inductors added together. For parallel connected inductors, the equivalent circuit inductance L_T is calculated differently.

The sum of the individual currents flowing through each inductor can be found using Kirchhoff's Current Law (KCL) where, $I_T = I_1 + I_2 + I_3$ and we know that the self-induced emf across an inductor is given as: $V = L\, di/dt$

Then by taking the values of the individual currents flowing through each inductor in our circuit above, and substituting the current i for $i_1 + i_2 + i_3$ the voltage across the parallel combination is given as:

$$V_{AB} = L_T \frac{d}{dt}\left(i_1 + i_2 + i_3 \right) = L_T \left(\frac{di_1}{dt} + \frac{di_2}{dt} + \frac{di_3}{dt} \right)$$

By substituting di/dt in the above equation with v/L gives:

$$V_{AB} = L_T \left(\frac{V}{L_1} + \frac{V}{L_2} + \frac{V}{L_3} \right)$$

We can reduce it to give a final expression for calculating the total inductance of a parallel circuit, and this is given as:

Parallel Inductor Equation

$$\frac{1}{L_T} = \frac{1}{L_1} + \frac{1}{L_2} + \frac{1}{L_3} \cdots\cdots + \frac{1}{L_N}$$

Here, like the calculations for parallel resistors, the reciprocal (1/Ln) value of the individual inductances are all added together instead of the inductances themselves. But again as with series connected inductances, the above equation only holds true when there is "NO" mutual inductance or magnetic coupling between two or more of the inductors, (they are magnetically isolated from each other). Where there is coupling between coils, the total inductance is also affected by the amount of coupling.

This method of calculation can be used for calculating any number of individual inductances connected together within a single parallel network. If however, there are only two individual parallel connected inductances then a much simpler and quicker formula can be used to find the total inductance value, and this is:

$$L_T = \frac{L_1 \times L_2}{L_1 + L_2}$$

One important point to remember about parallel inductive circuits, the total inductance (L_T) of any two or more inductors connected together in parallel will always be LESS than the value of the smallest inductance in the parallel branch.

Example No1

Three inductors of 60mH, 120mH and 75mH respectively, are connected together in a parallel combination with no mutual inductance between them. Calculate the total inductance of the parallel combination in millihenries.

$$\frac{1}{L_T} = \frac{1}{L_1} + \frac{1}{L_2} + \frac{1}{L_3}$$

$$\therefore L_T = \frac{1}{\dfrac{1}{L_1} + \dfrac{1}{L_2} + \dfrac{1}{L_3}} = \frac{1}{\dfrac{1}{60mH} + \dfrac{1}{120mH} + \dfrac{1}{75mH}}$$

$$L_T = \frac{1}{38.333} = 26mH$$

Mutually Coupled Parallel Inductors

When inductors are connected together in parallel so that the magnetic field of one links with the other, the effect of mutual inductance either increases or decreases the total inductance depending upon the amount of magnetic coupling that exists between the coils. The effect of this mutual inductance depends upon the distance apart of the coils and their orientation to each other.

Mutually connected parallel inductors can be classed as either "aiding" or "opposing" the total inductance with parallel aiding connected coils increasing the total equivalent inductance and parallel opposing coils decreasing the total equivalent inductance compared to coils that have zero mutual inductance.

Mutual coupled parallel coils can be shown as either connected in an aiding or opposing configuration by the use of polarity dots or polarity markers as shown below.

Parallel Aiding Inductors

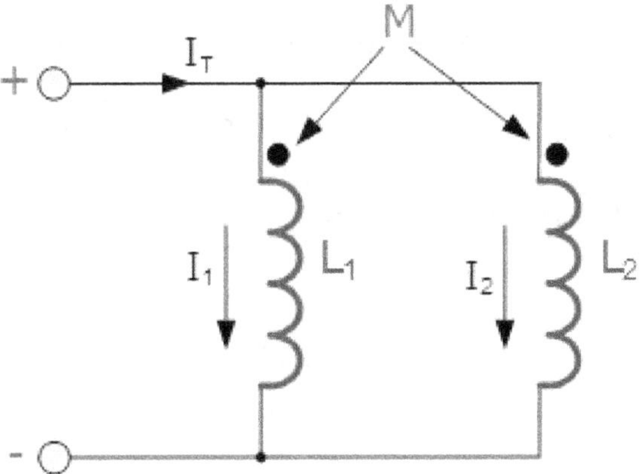

The voltage across the two parallel aiding inductors above must be equal since they are in parallel so the two currents, i_1 and i_2 must vary so that the voltage across them stays the same. Then the total inductance, L_T for two parallel aiding inductors is given as:

$$L_T = \frac{L_1 L_2 - M^2}{L_1 + L_2 - 2M}$$

Where: 2M represents the influence of coil L_1 on L_2 and likewise coil L_2 on L_1.

If the two inductances are equal and the magnetic coupling is perfect such as in a toroidal circuit, then the equivalent inductance of the two inductors in parallel is L as $L_T = L_1 = L_2 = M$. However, if the mutual inductance between them is zero, the equivalent inductance would be $L \div 2$ the same as for two self-induced inductors.

If one of the two coils was reversed with respect to the other, we would then have two parallel opposing inductors and the mutual inductance, M that exists between the two coils will have a cancelling effect on each coil instead of an aiding effect as shown below.

Parallel Opposing Inductors

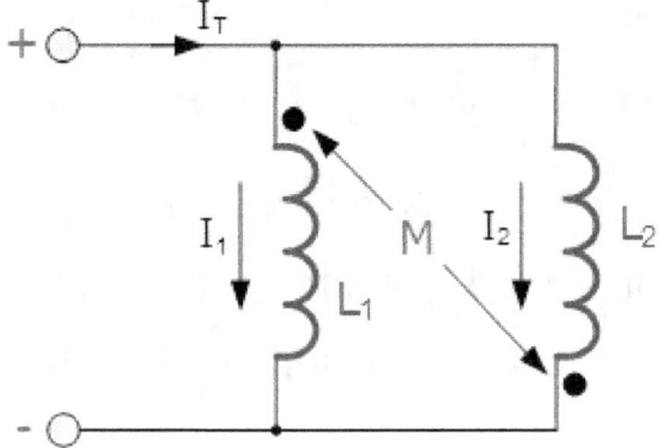

Then the total inductance, L_T for two parallel opposing inductors is given as:

$$L_T = \frac{L_1 L_2 - M^2}{L_1 + L_2 + 2M}$$

This time, if the two inductances are equal in value and the magnetic coupling is perfect between them, the equivalent inductance and also the self-induced emf across the inductors will be zero as the two inductors cancel each other out.

This is because as the two currents, i_1 and i_2 flow through each inductor in turn the total mutual flux generated between them is zero because the two fluxes produced by each inductor are both equal in magnitude but in opposite directions.

Then the two coils effectively become a short circuit to the flow of current in the circuit so the equivalent inductance, L_T becomes equal to (L ± M) ÷ 2.

Example No2

Two inductors whose self-inductances are of 75mH and 55mH respectively are connected together in parallel aiding. Their mutual inductance is given as 22.5mH. Calculate the total inductance.

$$L_T = \frac{L_1 \times L_2 - M^2}{L_1 + L_2 - 2M}$$

$$L_T = \frac{75\text{mH} \times 55\text{mH} - 22.5\text{mH}^2}{75\text{mH} + 55\text{mH} - 2 \times 22.5\text{mH}}$$

$$L_T = 42.6\text{mH}$$

Example No3

Calculate the equivalent inductance of the following inductive circuit.

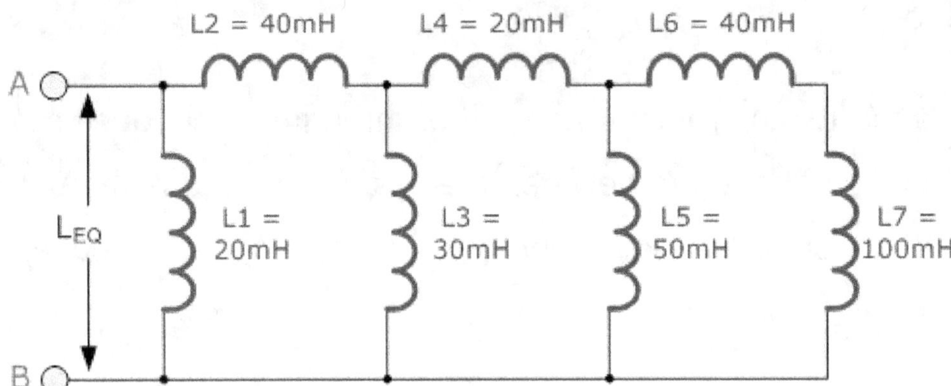

Calculate the first inductor branch L_A, (Inductor L_5 parallel with inductors L_6 and L_7)

$$L_A = \frac{L5 \times (L6 + L7)}{L5 + L6 + L7} = \frac{50mH \times (40mH + 100mH)}{50mH + 40mH + 100mH} = 36.8mH$$

Calculate the second inductor branch L_B, (Inductor L_3 in parallel with inductors L_4 and L_A)

$$L_B = \frac{L3 \times (L4 + L_A)}{L3 + L4 + L_A} = \frac{30mH \times (20mH + 36.8mH)}{30mH + 20mH + 36.8mH} = 19.6mH$$

Calculate the equivalent circuit inductance L_{EQ}, (Inductor L_1 in parallel with inductors L_2 and L_B)

$$L_{EQ} = \frac{L1 \times (L2 + L_B)}{L1 + L2 + L_B} = \frac{20mH \times (40mH + 19.6mH)}{20mH + 40mH + 19.6mH} = 15mH$$

Then the equivalent inductance for the above circuit was found to be: 15mH.

In a Brief

As with the resistor, inductors connected together in parallel have the same voltage, V across them. Also connecting together inductors in parallel decreases the effective inductance of the circuit with the equivalent inductance of "N" inductors connected in parallel being the reciprocal of the sum of the reciprocals of the individual inductances.

As with series connected inductors, mutually connected parallel inductors are classed as either "aiding" or "opposing" this total inductance depending whether the coils are cumulatively coupled (in the same direction) or differentially coupled (in opposite direction).

Thus far we have examined the inductor as a pure or ideal passive component. In the next   about Inductors, we will look at non-ideal inductors that have real world resistive coils producing the equivalent circuit of an inductor in series with a resistance and examine the time constant of such a circuit.

CHAPTER-6: LR SERIES CIRCUIT

The magnetic field that surrounds all coils, inductors, chokes, and transformers is created by an LR series circuit, which consists of an inductance in series with a resistance. The inductance and resistance of inductive coils and solenoids combine to provide a basic LR series circuit, rather than being entirely inductive devices.

An inductor's time constant indicate that it could not change instantly but would instead increase at a steady rate set by the self-induced back-emf generated by the inductive coil.

The flow of current, I through an inductor in an electrical circuit is, in other words, opposed. While this is absolutely true, we presupposed in the that the inductor was an ideal one with no resistance or capacitance attached to its coil windings.

However, "ALL" coils in the actual world, whether they are chokes, solenoids, relays, or any other coiled component, will always have some resistance, regardless of how little. This is due to the copper wire, which has a resistive value, being used to construct the actual coils of wire. Then, for practical purposes, we can think of our straightforward coil as a "Inductance," L, connected in series with a "Resistance," R.

Creating an LR Series Circuit, in other words. A resistor of resistance, R, and an inductor of inductance, L, are linked in series to form an LR series circuit.

The inductors coil's resistance ("R") is the DC resistive value of the wire turns or loops that make up the coil. Take the circuit in LR series shown below.

The LR Series Circuit

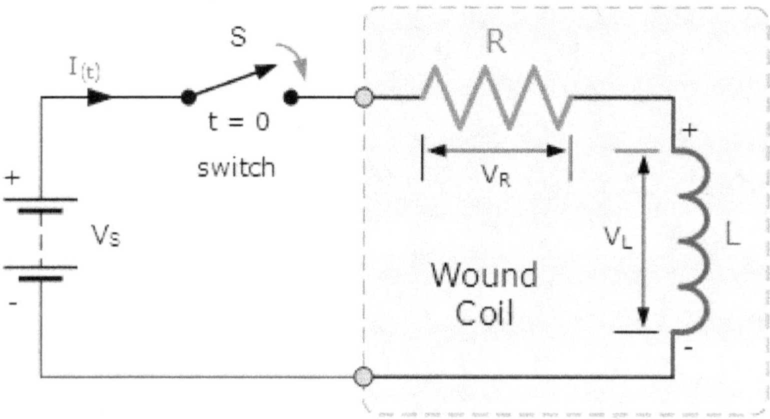

The above *LR series circuit* is connected across a constant voltage source, (the battery) and a switch. Assume that the switch, S is open until it is closed at a time t = 0, and then remains permanently closed producing a "step response" type voltage input. The current, i begins to flow through the circuit but does not rise rapidly to its maximum value of Imax as determined by the ratio of V / R (Ohms Law).

This limiting factor is due to the presence of the self-induced emf within the inductor as a result of the growth of magnetic flux, (Lenz's Law). After a time, the voltage source neutralizes the effect of the self-induced emf, the current flow becomes constant and the induced current and field are reduced to zero.

We can use Kirchhoff's Voltage Law, (KVL) to define the individual voltage drops that exist around the circuit and then hopefully use it to give us an expression for the flow of current.

Kirchhoff's voltage law (KVL) gives us:

$$V_{(t)} - (V_R + V_L) = 0$$

The voltage drop across the resistor, R is I*R (Ohms Law).

$$V_R = I \times R$$

The voltage drop across the inductor, L is by now our familiar expression L(di/dt)

$$V_L = L\frac{di}{dt}$$

Then the final expression for the individual voltage drops around the LR series circuit can be given as:

$$V_{(t)} = I \times R + L\frac{di}{dt}$$

It has been observed that the voltage drop across the resistor depends upon the current, i, while the voltage drop across the inductor depends upon the rate of change of the current, di/dt. When the current is equal to zero, (i = 0) at time t = 0 the above expression, which is also a first order differential equation, can be rewritten to give the value of the current at any instant of time as:

Expression for the Current in an LR Series Circuit

$$I_{(t)} = \frac{V}{R}\left(1 - e^{-Rt/L}\right) \ (A)$$

- Where:
- V is in Volts
- R is in Ohms
- L is in Henries
- t is in Seconds
- e is the base of the Natural Logarithm = 2.71828

The Time Constant, (τ) of the LR series circuit is given as L/R and in which V/R represents the final steady state current value after five time constant values. Once the current reaches this maximum steady state value at 5τ, the inductance of the coil has reduced to zero acting more like a short circuit and effectively removing it from the circuit.

So, the current flowing through the coil is limited only by the resistive element in Ohms of the coil's windings. A graphical representation of the current growth representing the voltage/time characteristics of the circuit can be presented as.

Transient Characteristics Curves

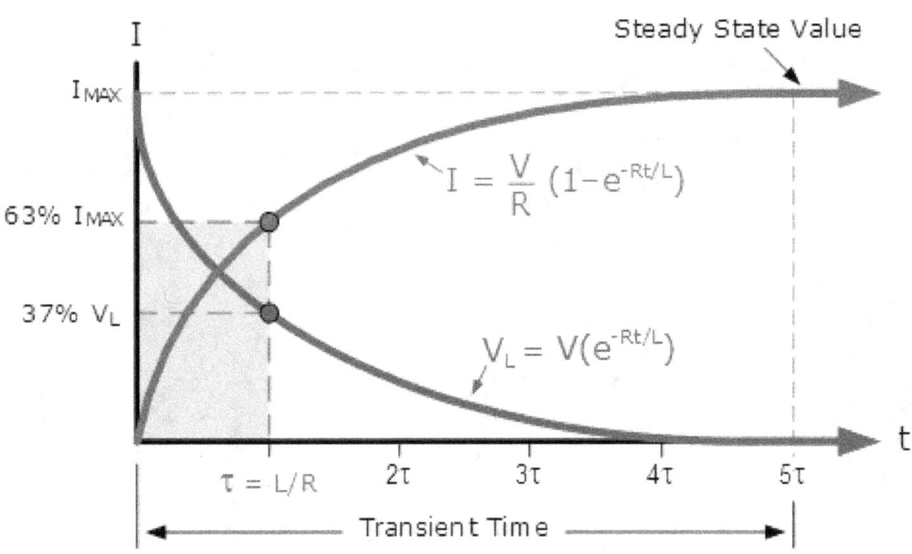

As the voltage drop across the resistor, V_R is equal to I*R (Ohms Law), it will have the same exponential growth and shape as the current. However, the voltage drop across the inductor, V_L will have a value equal to: $Ve^{(-Rt/L)}$. Then the voltage across the inductor, V_L will have an initial value equal to the battery voltage at time $t = 0$ or when the switch is first closed and then decays exponentially to zero as represented in the above curves.

The time required for the current flowing in the LR series circuit to reach its maximum steady state value is equivalent to about 5-time constants or 5τ.

This time constant τ, is measured by $\tau = L/R$, in seconds, where R is the value of the resistor in ohms and L is the value of the inductor in Henries. This then forms the basis of an RL charging circuit were 5τ can also be thought of as "5*(L/R)" or the *transient time* of the circuit.

The transient time of any inductive circuit is determined by the relationship between the inductance and the resistance. For example, for a fixed value resistance the larger the inductance the slower will be the transient time and therefore a longer time constant for the LR series circuit. Likewise, for a fixed value inductance the smaller the resistance value the longer the transient time.

However, for a fixed value inductance, by increasing the resistance value the transient time and therefore the time constant of the circuit becomes shorter. This is because as the resistance increases the circuit becomes more and more resistive as the value of the inductance becomes negligible compared to the resistance. If the value of the resistance is increased sufficiently large compared to the inductance the transient time would effectively be reduced to almost zero.

Example No1

A coil which has an inductance of 40mH and a resistance of 2Ω is connected together to form a LR series circuit. If they are connected to a 20V DC supply.

a). What will be the final steady state value of the current.

Steady State Current, $I = \dfrac{V}{R} = \dfrac{20}{2} = 10\,A$

b) What will be the time constant of the RL series circuit.

Time Constant, $\tau = \dfrac{L}{R} = \dfrac{0.04}{2} = 0.02\,s$ or $20\,mS$

c) What will be the transient time of the RL series circuit.

Transient Time, $5\tau = 5 \times 0.02\,s = 100\,mS$

d) What will be the value of the induced emf after 10ms.

Induced emf, $V_L = Ve^{(-Rt/L)} = 20\,e^{(-2\times 0.01/0.04)}$

$$V_L = 20 \times 0.6065 = 12.13\,V$$

e) What will be the value of the circuit current one time constant after the switch is closed.

Instantaneous Current, $I_{(t)} = \dfrac{V_S}{R}(1-e^{-Rt/L})$

The Time Constant, τ of the circuit was calculated in question b) as being 20ms. Then the circuit current at this time is given as:

$$I_{(t)} = \frac{20}{2}\left(1 - e^{-2\times 0.02/0.04}\right)$$

$$I_{(t)} = 10(1-0.368) = 6.32A$$

It has been observed that the answer for question (e) which gives a value of 6.32 Amps at one time constant, is equal to 63.2% of the final steady state current value of 10 Amps we calculated in question (a). This value of 63.2% or 0.632 x I_{MAX} also corresponds with the transient curves shown above.

Power in an LR Series Circuit

Then from above, the instantaneous rate at which the voltage source delivers power to the circuit is given as:

$$P = V \times I \quad \text{in Watts}$$

The instantaneous rate at which power is dissipated by the resistor in the form of heat is given as:

$$P = I^2 \times R \quad \text{in Watts}$$

The rate at which energy is stored in the inductor in the form of magnetic potential energy is given as:

$$P = Vi = Li\frac{di}{dt} \quad \text{in Watts}$$

Then we can find the total power in a RL series circuit by multiplying by i and is therefore:

$$P = i^2 R + Li\frac{di}{dt} \quad \text{(Watts)}$$

Where the first I^2R term represents the power dissipated by the resistor in heat, and the second term represents the power absorbed by the inductor, its magnetic energy.

Inductive Reactance

Since reactance is directly proportional to frequency, the frequency of the applied voltage affects the inductive reactance of a coil. Similar to the opposition to direct current (DC) in a resistance, inductive reactance is the characteristic of an inductive coil that resists the change in alternating current (AC) through it.

It has been observed that the behavior of inductors linked to DC supply and we know that ideally at this point, the value of the inductor's self-induced or back emf determines how quickly the current through it grows when a DC voltage is placed across it. Additionally, we are aware that the inductors current rises continuously until, after five-time constants, it reaches its maximum steady state condition.

Only the resistive portion of an inductive coil's windings in Ohms, which is defined by the voltage to current ratio, V/R, as we know from Ohms law, can limit the maximum current that can flow through an inductive coil. An inductor's current flow behaves substantially differently when an alternating or AC voltage is put across it than when a DC voltage is used.

A sinusoidal supply has the effect of producing a phase difference between the voltage and current waveforms. Now, in an AC circuit, the resistance to current flow through the coil windings relies on both the frequency of the AC waveform and the inductance of the coil.

The AC resistance, sometimes referred to as Impedance (Z), of the circuit determines the resistance to current flowing through a coil in an AC circuit. However, resistance is always related to DC circuits, therefore reactance is typically employed to differentiate between DC resistance and AC resistance.

Reactance is measured in Ohms, just like resistance, but is given the symbol X (an uppercase "X") to set it apart from simply resistive values. Since the component we are interested in is an inductor, the reactance of an inductor is therefore called "Inductive Reactance". In other words, an inductors electrical resistance when used in an AC circuit is called Inductive Reactance.

Inductive Reactance which is given the symbol X_L, is the property in an AC circuit which opposes the change in the current. In our   s about Capacitors in AC Circuits, we saw that in a purely capacitive circuit, the current I_C "LEADS" the voltage by 90°. In a purely inductive AC circuit, the exact opposite is true, the current I_L "LAGS" the applied voltage by 90°, or ($\pi/2$ rads).

AC Inductor Circuit

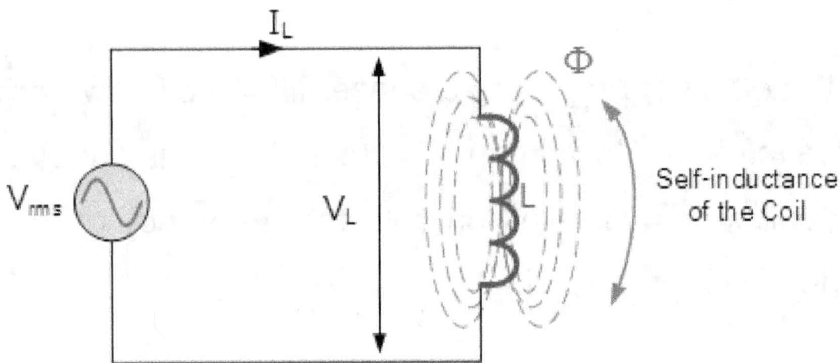

In the purely inductive circuit above, the inductor is connected directly across the AC supply voltage. As the supply voltage increases and decreases with the frequency, the self-induced back emf also increases and decreases in the coil with respect to this change.

We know that this self-induced emf is directly proportional to the rate of change of the current through the coil and is at its greatest as the supply voltage crosses over from its positive half cycle to its negative half cycle or vice versa at points, $0°$ and $180°$ along the sine wave.

So, the minimum rate of change of the voltage occurs when the AC sine wave crosses over at its maximum or minimum peak voltage level. At these positions in the cycle the maximum or minimum currents are flowing through the inductor circuit and this is shown below.

AC Inductor Phasor Diagram

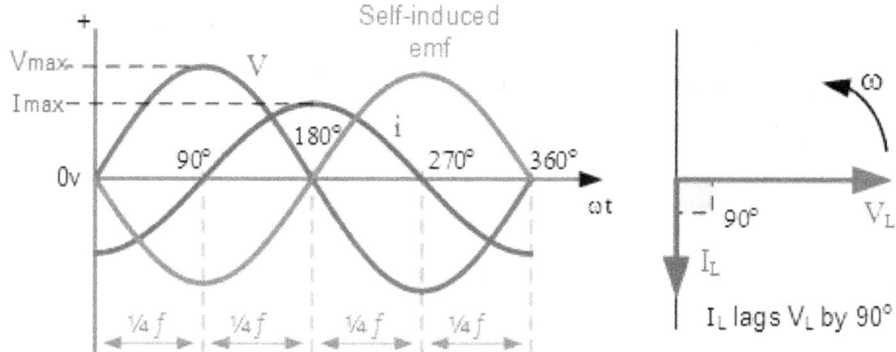

These voltage and current waveforms show that for a purely inductive circuit the current lags the voltage by 90°. Likewise, we can also say that the voltage leads the current by 90°. Either way the general expression is that the current lags as shown in the vector diagram. Here the current vector and the voltage vector are shown displaced by 90°. *The current lags the voltage.* This can also be written as, $V_L = 0°$ and $I_L = -90°$ with respect to the voltage, V_L. If the voltage waveform is classed as a sine wave, then the current, I_L can be classed as a negative cosine and we can define the value of the current at any point in time as being:

$$I_L = I_{max} \sin(\omega t - 90°)$$

Where: ω is in radians per second and t is in seconds.

Since the current always lags the voltage by 90° in a purely inductive circuit, we can find the phase of the current by knowing the phase of the voltage or vice versa. So if we know the value of V_L, then I_L must lag by 90°. Likewise, if we know the value of I_L then V_L must therefore lead by 90°.

Then this ratio of voltage to current in an inductive circuit will produce an equation that defines the Inductive Reactance, X_L of the coil.

CHAPTER-7: INDUCTIVE REACTANCE

$$X_L = \frac{V_L}{I_L} = \omega L \; (\Omega)$$

The preceding equation for inductive reactance can be rewritten in a more known form that uses the ordinary frequency of the supply rather than the angular frequency in radians, ω and this is given as:

$$X_L = 2\pi f L$$

Where: f is the Frequency and L is the Inductance of the Coil and $2\pi f = \omega$.

It is clear from the inductive reactance equation above that an increase in either frequency or inductance would result in an increase in overall inductive reactance.

The inductors' reactance would reach infinity as the frequency approached infinity, behaving as an open circuit. The inductors' reactance would, however, drop to zero when the frequency approached zero or DC, behaving as a short circuit.

Inductive reactance is therefore "proportional" to frequency, according to this. In other words, inductive reactance increases with frequency resulting in X_L being small at low frequencies and X_L being high at high frequencies and this demonstrated in the following graph:

Inductive Reactance against Frequency

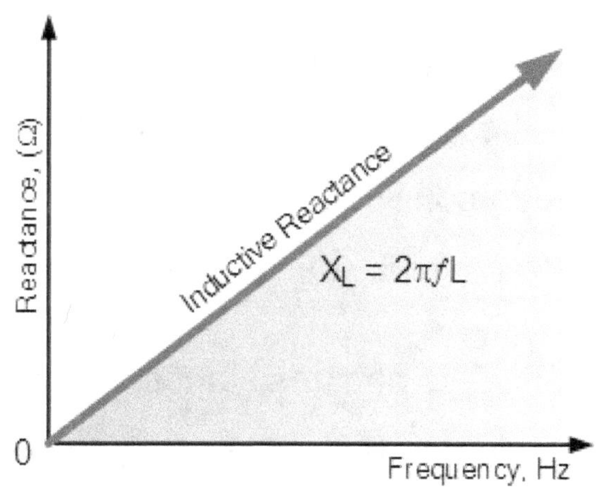

The slope shows that the "Inductive Reactance" of an inductor increases as the supply frequency across it increases.

Therefore Inductive Reactance is proportional to frequency giving:
($X_L \; \alpha \; f$)

Then we can see that at DC an inductor has zero reactance (short-circuit), at high frequencies an inductor has infinite reactance (open-circuit).

Inductive Reactance Example No1

A coil of inductance 150mH and zero resistance is connected across a 100V, 50Hz supply. Calculate the inductive reactance of the coil and the current flowing through it.

Inductive Reactance:
$$X_L = 2\pi f L = 2\pi \times 50 \times 0.15 = 47.12\Omega$$

Current:
$$I = \frac{V}{X_L} = \frac{100}{47.12} = 2.12A$$

AC Supply through an LR Series Circuit

Since all coils, relays, and solenoids have some resistance, regardless of how small it may be compared to the coil's wire turns, it is not possible to have an entirely inductive coil as we have been considering. Then, we can think of our straightforward coil as a resistance coupled with an inductance.

In an AC circuit that contains both inductance, L and resistance, R the voltage, V will be the phasor sum of the two component voltages, V_R and V_L. This means then that the current flowing through the coil will still lag the voltage, but by an amount less than 90° depending upon the values of V_R and V_L.

The new phase angle between the voltage and the current is known as the phase angle of the circuit and is given the Greek symbol phi, Φ.

To be able to produce a vector diagram of the relationship between the voltage and the current, a reference or common component must be found. In a series connected R-L circuit the current is common as the same current flows through each component. The vector of this reference quantity is generally drawn horizontally from left to right.

We know that the current and voltage in a resistive AC circuit are both "in-phase" and therefore vector, V_R is drawn superimposed to scale on the current or reference line.

We also know from above, that the current "lags" the voltage in a purely inductive circuit and therefore vector, V_L is drawn 90° in front of the current reference and to the same scale as V_R and this is shown below.

LR Series AC Circuit

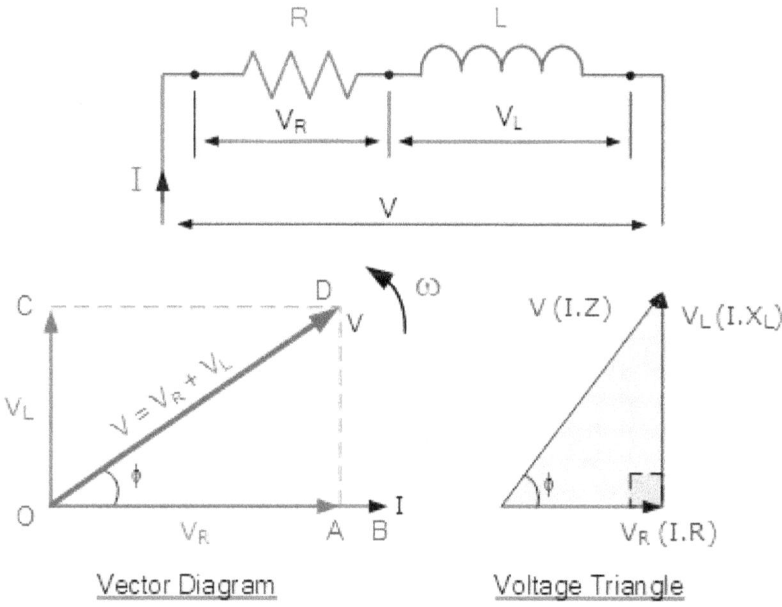

In the vector diagram above it can be seen that line OB represents the current reference line, line OA is the voltage of the resistive component and which is in-phase with the current. Line OC shows the inductive voltage which is 90º in front of the current, therefore it can be seen that the current lags the voltage by 90º. Line OD gives us the resultant or supply voltage across the circuit. The voltage triangle is derived from Pythagoras theorem and is given as:

$$V^2 = V_R^2 + V_L^2$$

$$V = \sqrt{V_R^2 + V_L^2}$$

$$\text{and} \quad \tan\Phi = \frac{V_L}{V_R}$$

$$\text{Since} \quad V_R = I \times R, \quad \text{and} \quad V_L = I \times X_L$$

$$V = \sqrt{(I \times R)^2 + (I \times X_L)^2}$$

$$\therefore I = \frac{V}{\sqrt{R^2 + X_L^2}} = \frac{V}{Z} \ (A)$$

In a DC circuit, the ratio of voltage to current is called resistance. However, in an AC circuit this ratio is known as Impedance, Z with units again in

Ohms. Impedance is the total resistance to current flow in an "AC circuit" containing both resistance and inductive reactance.

If we divide the sides of the voltage triangle above by the current, another triangle is obtained whose sides represent the resistance, reactance and impedance of the coil. This new triangle is called an "Impedance Triangle"

The Impedance Triangle

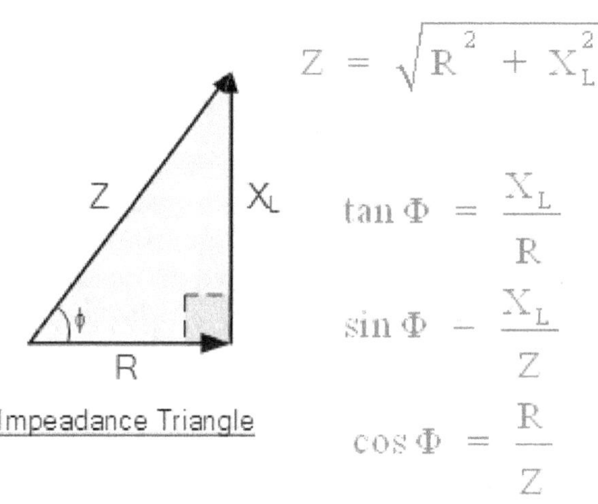

$$Z = \sqrt{R^2 + X_L^2}$$

$$\tan \Phi = \frac{X_L}{R}$$

$$\sin \Phi = \frac{X_L}{Z}$$

$$\cos \Phi = \frac{R}{Z}$$

Inductive Reactance Example No2

A solenoid coil has a resistance of 30 Ohms and an inductance of 0.5H. If the current flowing through the coil is 4 amps. Calculate,

a) The voltage of the supply if the frequency is 50Hz.

$$X_L = 2\pi f L$$
$$X_L = 2\pi.50.0.5$$
$$X_L = 157\,\Omega$$

$$Z = \sqrt{R^2 + X_L^2}$$
$$Z = \sqrt{30^2 + 157^2}$$
$$Z = 160\,\Omega$$

$$V = I \times Z = 4 \times 160 = 640V$$

b) The phase angle between the voltage and the current.

$$\tan \Phi = \frac{X_L}{R}$$

$$\therefore \Phi = \tan^{-1}\frac{X_L}{R} = \frac{157}{30} = 79.2^\circ \text{ lagging}$$

Power Triangle of an AC Inductor

The "Power Triangle" is another sort of triangle arrangement that can be used for an inductive circuit. Reactive power, also known as volt-amps reactive, or Var, is the term used to describe the power in an inductive circuit and is expressed in volt-amps.

The current lags the supply voltage by an angle of Φ° in an RL series AC circuit. In a purely inductive AC circuit, the current will be out-of-phase by a full 90° to the supply voltage. As such, the total reactive power consumed by the coil will be equal to zero as any consumed power is cancelled out by the generated self-induced emf power. In other words, the net power in watts consumed by a pure inductor at the end of one complete cycle is zero, as energy is both taken from the supply and returned to it.

The Reactive Power, (Q) of a coil can be given as: $I^2 \times X_L$ (similar to I^2R in a DC circuit). Then the three sides of a power triangle in an AC circuit are represented by apparent power, (S), real power, (P) and the reactive power, (Q) as shown.

Power Triangle

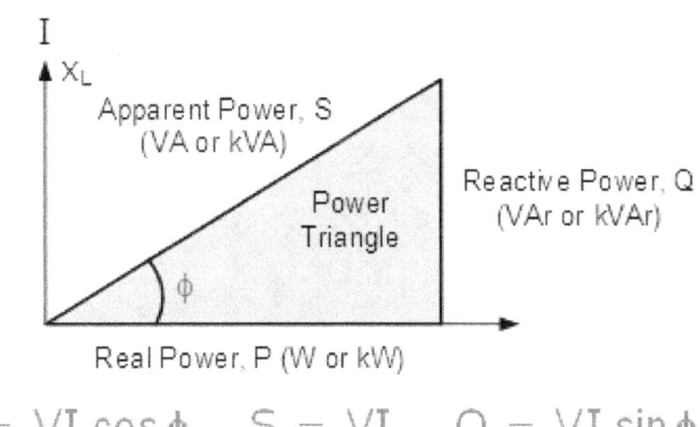

$$P = VI \cos\phi, \quad S = VI, \quad Q = VI \sin\phi$$

$$VA = \sqrt{P^2 + VAr^2}$$

Since the resistance of the windings creates an impedance, Z, an actual inductor or coil will consume power in watts.

www.ingramcontent.com/pod-product-compliance
Lightning Source LLC
Chambersburg PA
CBHW080849160726
47999CB00009B/3049